From R. C. Davis

STATEMENT

OF THE

OBJECT AND ORGANIZATION

OF THE

AMERICAN

Geographical and Statistical Society;

WITH

A COPY OF ITS CHARTER, BY-LAWS,

FORMATION OF SECTIONS,

AND

LIST OF MEMBERS.

NEW YORK:
BAKER & GODWIN, PRINTERS.
CORNER NASSAU AND SPRUCE STREETS.
1857.

STATEMENT

OF THE

OBJECT AND ORGANIZATION

OF THE

AMERICAN

Geographical and Statistical Society;

WITH

A COPY OF ITS CHARTER, BY-LAWS,

FORMATION OF SECTIONS,

AND

LIST OF MEMBERS

NEW YORK:
BAKER & GODWIN, PRINTERS,
CORNER NASSAU AND SPRUCE STREETS.
1857.

The following is the list of officers for the year 1857:

PRESIDENT,

FRANCIS L. HAWKS, D. D., LL. D.

VICE-PRESIDENTS,

HENRY GRINNELL, ARCHIBALD RUSSELL,
ALEXANDER W. BRADFORD.

FOREIGN CORRESPONDING SECRETARY,

JOHN JAY.

DOMESTIC CORRESPONDING SECRETARY,

FREDERICK A. CONKLING.

RECORDING SECRETARY,

J. C. ADAMSON.

ASSISTANT RECORDING SECRETARY,

SAMUEL P. DINSMORE.

TREASURER,

RIDLEY WATTS.

LIBRARIAN,

MARSHALL LEFFERTS.

COUNCIL,

HIRAM BARNEY, GEORGE FOLSOM,
J. CALVIN SMITH, R. A. WITTHAUS,
HENRY V. POOR, ABRAM S. HEWITT,
M. DUDLEY BEAN, HENRY E. PIERREPONT,
EGBERT L. VIELE.

STANDING COMMITTEES OF THE COUNCIL,

FOR THE YEAR 1857.

On Topics and Proceedings,

Messrs. CONKLING, BARNEY, and VIELE.

On Printing,

Messrs. POOR, BRADFORD, and VIELE.

On the Library and Finance,

Messrs. LEFFERTS, BEAN, and HEWITT.

On Recent Discoveries and Publications,

Messrs. WITTHAUS, FOLSOM, and SMITH.

The Chairman of the Council and Recording Secretary are *ex-officio* Members of the several Standing Committees.

Object and Plans of the Society.

The object of the Society is the advancement of Geographical and Statistical Science, by the collection and diffusion of knowledge in those branches.

Among the measures at present contemplated for this purpose, are the following:

1. A collection of the most valuable maps and books of reference on Geography and Statistics, from all countries, to be deposited and kept for public use, under proper regulations.

2. Stated meetings on Thursday of every week (July, August, and September excepted), open to the public, for the reading of valuable papers on Geographical and Statistical subjects, collected from all quarters, with conversational discussions, personal narratives of explorers and travelers, &c.

3. A Bulletin of the Society's transactions and collections, published periodically, furnished to all members, and sent in exchange to kindred institutions in all countries.

4. To obtain, through the kind co-operation of the foreign Ministers and Consuls resident here, the latest, completest, and most authentic information, publications, and public documents, from their respective countries.

5. By correspondence with Missionaries, Officers of the Army and Navy, and others, to increase the extent and accuracy of Geographical and Statistical knowledge.

6. Originating and assisting in explorations of undescribed regions, and in voyages of discovery.

Contributions in aid of these several plans, will be thankfully received and acknowledged, whether of maps and works of reference, papers and communications on topics of interest, recent and reliable information concerning any part of the world, or means to defray the unavoidable expenses of the Society.

CHARTER.

AN ACT *to incorporate the American Geographical and Statistical Society,* passed April 13, 1854.

The People of the State of New York, Represented in Senate and Assembly, do enact as follows:

§ 1. George Bancroft, Henry Grinnell, Francis L. Hawks, John C. Zimmerman, Archibald Russell, Joshua Leavitt, William C. H. Waddell, Ridley Watts, S. DeWitt Bloodgood, M. Dudley Bean, Hiram Barney, Alexander J. Cotheal, Luther B. Wyman, John Jay, J. Calvin Smith, Henry V. Poor, Cambridge Livingston, Edmund Blunt, Alexander W. Bradford, and their associates, who are now or may become hereafter associated for the purposes of this act, are hereby constituted a body corporate by the name of the American Geographical and Statistical Society, for the purpose of collecting and diffusing geographical and statistical information.

§ 2. For the purposes aforesaid the said Society shall possess the general powers and privileges, and be subject to the general liabilities contained in the third title of the eighteenth chapter of the first part of the Revised Statutes, so far as the same may be applicable and may not have been modified or repealed; but the real and personal estate which the said Society shall be authorized to take, hold, and convey, over and above its library, and maps, charts, instruments, and collections, shall not at any time exceed an amount the clear yearly income of which shall be ten thousand dollars.

§ 3. The officers of the said Society shall be a President, three Vice-Presidents, a Corresponding Secretary, a Recording Secretary, a Librarian, and Treasurer, and such other officers as may from time to time be provided for by the By-Laws of the said Society.

§ 4. The said Society, for fixing the terms of admission of its members for the government of the same, for changing and altering the officers above named, and for the general regualtion and management of its transactions and affairs, shall have power to form a code of By-Laws not inconsistent with the laws of this State or of the United States; which code, when formed and adopted at a regular meeting, shall, until modified or rescinded, be equally binding as this act upon the said Society, its officers and its members

§ 5. The Legislature may at any time alter or repeal this act.

§ 6. This act to take effect immediately.

STATE OF NEW YORK,
Secretary's Office.

I have compared the preceding with the original law on file in this office, and hereby certify the same to be a correct transcript therefrom and of the whole of said original law.

[L. S.] Given under my hand and seal of office, at the city of Albany, this thirteenth day of April, one thousand eight hundred and fifty-four.

A. G. JOHNSTON,
Dep. Sec'y of State.

BY-LAWS

OF THE

American Geographical and Statistical Society.

MEMBERS.

I. The Society shall consist of *Resident*, *Corresponding*, *Honorary*, *and Ex-officio Members*. *Corresponding Members* shall be limited to those who have advanced, by their publications or services, the Science of Geography or Statistics. *Honorary Members*, to those who have distinguished themselves in the Science of Geography or Statistics, but at no time to exceed twenty persons. *Ex-officio Members* shall be, the Foreign Ministers and Consuls resident in the United States, and American Ministers and Consuls resident abroad.

II. Persons desirous of joining the Society as Resident Members, shall be nominated to the Society or the Council in writing, and if recommended by the Council, shall, when elected by the Society, on the payment of the initiation fee of five dollars, become members.

Honorary or Corresponding Members may be elected by the Society, on the recommendation of the Council.

Ex-officio Members shall be enrolled, on their own request, or by vote of the Council; and any Member may, by a unanimous vote of the Society, be elected without reference to the Council.

ANNUAL DUES.

II. The Annual Dues of Resident Members shall be five dollars, payable in advance.

Should any Resident Member refuse to pay the Annual Dues when called upon, or neglect the payment for two years successively, the Council may erase his name from the list of Members.

The Council may at any time, by resolution, remit the initiation and annual dues of any person.

Any member, before the payment of his initiation fee, may become a Life Member upon the payment of the sum of fifty dollars, which shall be deemed to be inclusive of the initiation fee, and all future dues; and any member, subsequently to the payment of his initiation fee, may become a Life Member upon the payment of the sum of fifty dollars, which shall be in lieu of all subsequent annual dues.

OFFICERS.

III. The officers of the Society are—

A President,
Three Vice-Presidents,
A Foreign Corresponding Secretary,
A Domestic Corresponding Secretary,
A Recording Secretary,
An Assistant Recording Secretary,
A Treasurer, and
A Librarian.

They shall be elected annually by ballot, and shall hold their offices respectively for one year, and until others shall be chosen in their places.

The officers of the Society shall be members of the Council *ex-officio.*

ANNUAL MEETING.

IV. The Society shall hold its Annual Meeting on the first Thursday of December, yearly, when the officers and nine members of Council shall be chosen by ballot, a majority of votes being necessary for a choice on the first balloting; but where no choice is effected to any office on the first trial, a further balloting shall take place, at which a plurality of votes shall elect. At such Annual Meeting a written report shall be submitted by the Council of the proceedings of the Society during the past year, and also a written report by the Corresponding Secretaries, Librarian, and Secretaries, in their respective departments.

MONTHLY AND SPECIAL MEETINGS.

V. The Society shall meet regularly for the transaction of business, at its room, on the first Thursday in every month, unless otherwise specially ordered. But the President, or in his absence any of the Vice-Presidents, may, and upon the written request of any fifteen members shall, call a Special Meeting, giving three days' notice thereof in two public newspapers printed in the city of New York, and a special notice of at least three days, sent through the Post-office, to each Resident Member.

ORDER OF BUSINESS.

VI. At the regular meetings of the Society, the following shall be the order of business:

1. The reading of the Minutes of the last meeting.
2. Report of the Council.
3. Reports and Communications from officers of the Society.
4. Reports from Committees.
5. Miscellaneous business.
6. Papers read and Addresses delivered before the Society.

QUORUM.

VII. At all meetings of the Society, eight members shall constitute a quorum for the transaction of business.

PRESIDING OFFICER.

VIII. The President, or in his absence one of the Vice-Presidents, or in their absence a Chairman *pro tempore*, shall preside at all the meetings of the Society, and shall have a casting vote. He shall preserve order, and shall decide all questions of order, subject to an appeal to the Society. He shall also appoint all committees authorized by the Society, unless otherwise specially ordered.

CORRESPONDING SECRETARIES.

IX. The Corresponding Secretaries shall conduct the general correspondence of the Society. They shall have the custody of all letters and communications to the Society, excepting papers read and addresses delivered before the same, which shall be deposited in the library. They shall, at every meeting of the Society, read such letters and communications as they may have received; they shall prepare all letters to be written in connection with the business or objects of the Society, and transmit the same; but the Society may appoint a Committee to prepare a letter or letters on any special occasion. They shall keep, in suitable books to be provided for that purpose, true copies of all letters written on behalf of the Society; and shall carefully preserve the originals of all letters and communications received.

The duties of the Foreign Corresponding Secretary shall be limited to the correspondence with individuals or associate bodies in foreign countries; and those of the Domestic Corresponding Secretary shall, in like manner, be confined to the United States; except that, in the absence of either of these officers, or during a vacancy in either office, its duties shall be performed by the remaining incumbent, until such absence shall terminate, or the vacancy be supplied.

RECORDING SECRETARY.

X. The Recording Secretary shall have the custody of the Seal, Charter, By-Laws, and Records of the Society. He shall, under the direction of the President, or either of the Vice-Presidents, give due notice of the time and place of all meetings of the Society, and attend the same. He shall keep fair and accurate records of all the proceedings and orders of the Society; and shall give notice to the several officers, and to the Trustees and Committees, of all votes, orders, resolves, and proceedings of the Society affecting them, or appertaining to their respective duties.

LIBRARIAN.

XI. The Librarian, under the direction of the Board of Trustees, shall have the charge and superintendence of the rooms, and the custody and arrangement of the books, maps, and other articles belonging to the Library and Collections of the Society. He shall acknowledge the receipt of donations to the Society in his department.

TREASURER.

XII. The Treasurer shalt collect and keep the funds and securities of the Society, and shall disburse and dispose of the same under the direction of the Council.

COUNCIL.

XIII. The Council shall have the control and management of the affairs, property, and funds of the Society.

Five members shall be a quorum for the transaction of business.

They shall have power to fill any vacancy that may occur among the Officers or Council of the Society.

They shall make their own by-laws, which shall not be inconsistent with these by-laws.

They shall meet at least once in every month, for the transaction of business, except during the months of July, August, and September.

If any member of the Council shall be absent from its meetings for three successive months, without reasons satisfactory to the Council, his place may be declared vacant by a vote of the Council; but this rule shall not apply to the President or Vice-Presidents.

Any officer or member of the Council may be removed from his place by the Council, for cause, by a vote of two-thirds of the members present at a meeting, provided such two-thirds shall be equal in number to a majority of the whole Council.

The Council shall not make any contract involving a liability exceeding the sum of five hundred dollars, nor make any sale or disposition of the property of the Society exceeding in value that sum, without an approving vote of the Society at a regular meeting.

No debt shall be contracted whereby the aggregate debts of the Society shall be increased more than $1,000 above the unappropriated funds at the time in the treasury, and no debt shall be incurred until it shall have been approved by the Council.

SECTIONS.

XIV. It shall be the duty of the Council to distribute into departments the various subjects which properly belong to Geographical and Statistical science, each of which departments shall be confided to a section of the Society; and the Council shall appoint annually a suitable person as chairman of each section, and prescribe from time to time such rules and regulations for the government of the sections as may to the Council seem advisable.

Any member of the Society may attach himself to such section or sections, as he may wish, by giving notice of his desire to the Recording Secretary, or to the chairman of such section.

It shall be the duty of the chairman of each section to induce members of the Society to become associated with such sections as they take an interest in, that by mutual labor and co-operation they may advance the cause of science.

No debt nor expenditure shall be incurred by any section without a previous appropriation by the Council; but the members of any section may raise money for their own objects, and expend the same without the authority of the Council.

The Council shall regulate the manner and order in which the sections shall report to the Society.

ALTERATION OF BY-LAWS.

XV. No alteration in the By-Laws of the Society shall be made, unless such alteration shall have been openly proposed at a previous meeting and entered on the Minutes, with the name of the member proposing the same, and unless it shall be adopted by a vote of two-thirds of the members present at a regular meeting of the Society.

By-Laws of the Council.

1. The Council shall elect its own Chairman and Secretary.

2. Stated meetings of the Council shall be held at the rooms of the Society every Thursday, excepting during the months of July, August, and September, and at such other times as the Council may determine.

3. Special meetings may be called at any time, by the Secretary, by direction of the Chairman, or upon the written request of five members of the Council.

4. The order of business shall be as follows:
    1. Reading of the Minutes.
    2. Report of Treasurer.
    3. Report of the Librarian and Superintendent.
    4. Reports of Committees.
    5. Consideration of Nominations.
    6. Unfinished business.
    7. Miscellaneous business.
    8. Papers and Communications to the Council may be read at any time, at the discretion of the Chairman.

5. There shall be four Standing Committees, to be called:
    1. Committee on Topics and Proceedings.
    2. Committee on Printing.
    3. Committee on the Library and Finance.
    4. Committee on Recent Discoveries and Publications.

6. The Committee on Topics and Proceedings shall procure suitable papers to be read before the Society, or before the Council, and recommend topics of discussion at the several meetings.

7. The Committee on Printing shall prepare the Bulletin, and superintend the printing and publication thereof. They shall also prepare for printing, all blanks, diplomas, notices, and other documents required for the transaction of the Society's business.

8. The Committee on the Library and Finance shall have charge of the property of the Society; shall make the requisite arrangements for the preservation thereof; arrange the hours during which the rooms shall be opened; shall provide for the accommodation of all meetings of, and lectures before, the Society or Council.

9. The Committee on Recent Discoveries and Publications shall procure and present to the Society or Council, from time to time, notices, abstracts, and reviews of recent books, reports, tables, maps, and other documents relating to Geography or Statistics.

10. The Standing Committees shall each consist of three members, to which shall be added the Chairman and Recording Secretary as ex-officio members. Two members shall constitute a quorum.

11. It shall be the duty of the Chairman to call the attention of the Council to the continued absence of any member for three successive months, and to submit to the Council the propriety of acting under the By-Law of the Society provided for such cases.

12. It shall be the duty of the Secretary to keep a record of the names of those present at each meeting of the Council.

13. Special meetings of the Council may always be held after meetings of the Society, without further notice.

14. These By-Laws may be temporarily suspended, by the vote of the majority of the members present at any meeting; but no amendment or addition thereto shall be made, except by a vote of a majority of the members of the Council, at a meeting subsequent to one at which notice of such amendment or addition shall have been given.

Foreign Societies with whom the American Geographical and Statistical Society is in Correspondence.

1. The Royal Geographical Society of London.
2. The Royal Society of London.
3. The Statistical Society of London.
4. The Institute of France, at Paris.
5. The French Society of Universal Statistics, at Paris.
6. The French Society of Geography, Natural History, and Agriculture, at Paris.
7. The National Society of Agriculture, Natural History, and Useful Arts, at Lyons.
8. The Society of Sciences, Agriculture, and Arts of the Lower Rhine, at Strasburg.
9. The Imperial Academy of Sciences, at Vienna.
10. The Royal Bohemian Society of Sciences, at Prague.
11. The Imperial Academy of Sciences, at St. Petersburg.
12. The Royal Academy of Science and Fine Arts, at Brussels.
13. The Royal Swedish Academy of Science, at Stockholm.
14. The Royal Norwegian University of Christiania.
15. The Royal Danish Society of Sciences, at Copenhagen.
16. The Royal Bavarian Society of Sciences, at Munich.
17. The Royal Society of Sciences, at Göttingen.
18. The Geographical Society, at Berlin.
19. The Botanical Society of Netherlands, at Leyden.
20. The Society of Swiss Antiquities, at Basle.
21. The Society of Swiss Antiquities, at Zurich.
22. The Commercial Library, at Hamburg.
23. The Academy of Science, Letters and Fine Arts, at Genoa.
24. The Practical Academy of Science and Belles-Lettres, at Naples.
25. The Astronomical Observatory of the Roman College, at Rome.
26. The Senckenberg Society of Natural History, at Frankfort-on-the-Maine.
27. The Royal Society, at Edinburgh.
28. The Board of Admiralty, at London.
29. The Geographical Society of Bombay.
30. The Asiatic Society of China, at Hong-Kong.
31. Royal Society of Van Diemen's Land, at Hobartstown.
32. The Society Auxiliary to National Industry, at Rio Janeiro.
33. The Society for the Advancement of Practical Economy, at Havana.
34. The Mexican Society of Geography and Statistics, at Mexico.

FORMATION OF SECTIONS.

DIVISIONS AND SUBDIVISIONS

OF THE

Several Subjects of Geographical and Statistical Investigation.

DEPARTMENT OF PHYSICAL GEOGRAPHY.

Geology.

E. L. VIELE, *Chairman*, 27 Amity Street.

General Geology of North and South America.
Structure and characteristics of the mountain systems.
Alluvial and other deposits.
Geographical distribution of metals and coal.
Paleontology.
Recent geological explorations and discoveries.
Special Geology of the several States and Territories.

Topography.

H. V. POOR, *Chairman*, 83 St. Mark's Place.

General Topography of North and South America.
Water-sheds.
Valleys of drainage.
Plateaus and basins.
Topography of the several States and Territories in detail.
Topography of the islands in the Atlantic and Pacific Oceans, and in the Gulf of Mexico.
Maps, charts, geodetic surveys, recent surveys, and their results.

Hydrology.

REV. DR. HAWKS, *Chairman*, 64 East 21st Street.

Seas, bays, and inland waters.
Currents.
Rivers; their source, character, and commercial importance.
Rivers; improvement of.
Canals; new, old, and projected.
Results of soundings.
Peculiarities of the American sea-coast.
Character of Harbors; Improvement of Harbors.
Water supply to cities and towns.

Zoology and Botany.

F. A. CONKLING, *Chairman*, 103 Tenth St.

Geographical distribution of men and the inferior animals in America.
Animals of domestic and commercial value.
Their comparative increase or decrease, in various localities.
Distribution, growth, and reproduction of plants.
Plants; quantity and quality of, commercially considered.
Cereals.

Meteorology.

H. E. Pierrepont, *Chairman*, box 24, P. O.

Geographical distribution of climate, and climatic causes.

Comparative climate of the different States and Territories.

Effects of climate on animal life.

Rain, snow, winds, clouds, temperature, mists, hurricanes, storms.

Meteors, electricity, magnetism.

Astronomy.

Rev. Dr. Adamson, *Chairman*, Recording Secretary.

Results and character of astronomical observations in America.

Recent discoveries.

History and Progress of Geographical Science at Home and Abroad.

Hon. G. Folsom, *Chairman*, 212 2d Avenue.

Discoveries.

Travels.

Geographical publications.

DEPARTMENT OF STATISTICS.

Commerce and Navigation.

Ridley Watts, *Chairman*, 27 East 20th St.

Ships and steamers.

Railways, rivers, canals, roads.

Imports and exports, foreign and coastwise.

Inland trade.

Marine and Fire Insurance.

Mines and Manufactures.

Marshall Lefferts, *Chairman*, 428 Fourth Street.

Minerals, commercially valuable.

Minerals, production of.

Late inventions in machines and machinery.

Economical processes in manufactures.

Agriculture.

John Jay, *Chairman*, 5th Avenue, near 35th Street.

Amount of arable and of cultivated and uncultivated lands in America.

Comparative products.

Staple productions.

Domestic animals.

Revenue from agriculture.

Finance.

R. A. Witthaus, *Chairman*, 86 2d Avenue.

National wealth and indebtedness.

Wealth and debts, State and municipal.

Moral Condition.

Arch'd Russell, *Chairman*, 45 10th St.

Ecclesiastical statistics; expense of supporting religion.

Eleemosynary statistics; expense of charities, public and private.

Educational statistics; expense of universities, colleges, schools, &c.

Criminal statistics.

Prison discipline; expense of, and results.

Vital Statistics.

M. D. Bean, *Chairman*, 48 West 14th St.

Medical statistics.

Population.

Life insurance and annuities.

Military and naval statistics, as relating to human life.

Sanitary regulations.

Political Statistics.

S. P. Dinsmore, *Chairman*, S. E. cor. 14th Street and 7th Avenue.

Elective franchise; number of voters, native, naturalized.

Immigration; whence, resources of immigrants, where settled, &c.

Fiscal; revenue, increase or decrease—taxes, direct or indirect.

Subjects of taxation.

Cost of supporting General Government; comparative cost in the several States.

HONORARY MEMBERS.

[Number limited to twenty.]

BACHE, PROF. A. D.	Washington.
BARTH, DR. HENRY	Gotha.
DE LA ROQUETTE, M.	Paris.
FREMONT, JOHN CHARLES . . .	New York.
HUMBOLDT, BARON ALEX. VON . . .	Berlin.
KANE,* DR. ELISHA K.	
KEILHAU, PROF. M. B.	Russia.
LAYARD, AUSTEN H.	London.
MAURY, LIEUT. M. F.	Washington.
MIDDENDORFF, PROF. VON . . .	St. Petersburg.
MURCHISON, SIR RODERICK I. . . .	London.
PETERMANN, DR. AND PROF. AUGUSTUS	Gotha.
QUETELET, LAMBERT A. J. . . .	Brussels.
RAWLINSON, LIEUT. COL. SIR HENRY C.	London.
RITTER, PROF. CARL VON . . .	Berlin.
ROBINSON, REV. EDWARD, D. D. . .	New York.
SIMPSON, SIR GEORGE	Montreal.
STRUVE, PROF. OTTO VON . . .	St. Petersburg.

* Deceased.

CORRESPONDING MEMBERS.

Abert, Col. J. J.	Washington.
Alexander, John H.	Baltimore.
Chaix, Prof. Paul	Geneva, Switzerland.
Cook, Charles W.	San Francisco.
Cornell, Miss S. S.	New York.
Dickens, Asbury	Washington.
Duer, Lieut. J. K.	U. S. N.
Duncan, Wm. Henry	Hanover, N. H.
Force, Peter	Washington.
Gilman, Daniel C.	
Gilpin, Hon. Henry D.	Philadelphia.
Green, Gen. Thomas J. . . .	Washington.
Gulick, Dr. L. H.	Micronesian Islands.
Guyot, Prof. A.	Princeton, N. J.
Kennedy, Hon. J. P.	Baltimore.
Leavenworth, Hon. E. W.	Syracuse, N. Y.
Livingston, Rev. David, D.D. . .	London.
Martin, R. Montgomery	London.
Mason, Hon. Charles	Washington.
Meehan, J. S.	Washington.
McClelland, Hon. Robert	Washington.
Negri, Christoforo	Turin.
Norman, Benjamin M.	New Orleans.
Olrick, J. C., Royal Inspector . .	North Greenland.
Palmer, Aaron H.	Washington.
Rogers, Prof. Henry D.	Boston.
Schade, Louis	Washington.
Shaw, Norton, Roy. Geog. Soc., . .	London.
Seward, Hon. Wm. H.	Auburn, N. Y.
Seymour, Hon. Horatio	Albany.
Smith, Edward R.	Washington.
Somerville, Mrs.	England.
Stevens, Gov. Isaac I.	Olympia, W. T.
Ward, Col. Thomas W.	Tarrytown, N. Y.
Waring, George E., Jr..	Westchester.
Watts, Burghall G.	Carthagena, N. Granada.

RESIDENT MEMBERS.

*** *Those having* L. M. *preceding their names have compounded for life. The List includes all elections up to June* 1, 1857.

1856........	Abbott, Gorham D.,	7 *Union Place.*
1856........	Abrams, John D.,	233 *West Nineteenth St.*
1857........	Adamson, James C., D. D.,	128 *Christopher St.*
1857........	Addoms, Charles,	75 *East Twenty-third St.*
1855........	Allen, William H,	428 *Fourth St.*
1853........	Alsop, Joseph W.,	33 *Washington Square.*
1856........	Alvord, Alonso A.,	10 *West Thirty-first St.*
1856........	Ames, Isaac,	119 *West Nineteenth St.*
1857........	Anderson, Henry J.,	62 *West Twentieth St.*
1856........	Annan, William C.,	94 *Wall St.*
1856........	Anthony, Jr., Jacob,	16 *Nassau St.*
1854........	Appleton, John A.,	36 *West Twenty-third St.*
1856........	Appleton, Wm. H.,	139 *Macdougal St.*
1852........	Arpin, Paul,	44 *Irving Place.*
1856........	Aspinwall, Wm. H.,	33 *University Place.*
1853........	Astor, John Jacob,	100 *Tenth St.*
1856........	Averill, Augustin,	7 *East Seventeenth St.*
1856........	Averill, J. Otis.	
1852....L. M.	Bancroft, George,	17 *West Twenty-first St.*
1853........	Baird, Rev. Robert, D. D.,	11 *Clinton St., Brooklyn.*
1856........	Baker, David F.,	*Pine St.*
1853........	Baker, Peter C.,	*Tribune Buildings.*
1856........	Ball, Henry,	717 *Houston St.*
1857........	Baldwin George D.,	222 *Wooster St.*
1856........	Baldwin, Simeon,	48 *East Twenty-fifth St.*
1856........	Ballard, Loomis,	123 *Maiden Lane.*
1856........	Ballow, William B.,	24 *East Twenty-Second St.*
1856........	Barker, B. Fordyce, M. D.,	70 *Union Place.*
1852....L. M.	Barney, Hiram,	266 *Fourth St.*

RESIDENT MEMBERS.

1856........Barstow, Caleb, ... 7 *Wall St.*
1856........Bayard, Edward, M. D., . . . 6 *West Fourteenth St.*
1856........Bassett, A. E.
1852........Bean, M. Dudley,. . . . . 48 *West Fourteenth St.*
1856........Beach, Oakley, ... 61 *Cortlandt St.*
1856........Beadle, Edward L., . . . 42 *Bleecker St.*
1857........Beakley, George, M. D., . . . 251 *Fourth Avenue.*
1856........Beakley, Jacob, M. D. . . . 251 *Fourth Avenue.*
1856........Beebee, Samuel J., ... 47 *Wall St.*
1857........Beekman, James W., . . . 234 *Fourth Avenue.*
1856........Berry, Richard, ... 65 *East Seventeenth St.*
1853........Bird, George, ... 1 *East Seventeenth St.*
1856........Blakeman, Birdsey, ... 115 *Nassau St.*
1856........Black, William, ... 247 *Broadway.*
1856........Bliss, Dallett, ... 27 *Old Slip.*
1856........Bleecker, James, ... 7 *Broad Street.*
1852........Bloodgood, Samuel DeWitt, . . 29 *East Twenty-eighth St.*
1852........Blunt, Edmund, ... 27 *Munroe Place.*
1852........Blunt, George W., ... 139 *West Twelfth St.*
1856........Bogart, Alexander J., . . . 315 *Pacific St., Brooklyn.*
1855........Bogert, Henry K., ... 49 *William St.*
1855........Bonney, Benj. W., . . . 38 *Wall St.*
1856........Booth, Wm. A., ... 66 *West Fourteenth St.*
1857........Booth, William T., . . . 66 *West Fourteenth St.*
1852....... Bradford, Alex. W., . . . 140 *Ninth Avenue.*
1852........Brevoort, J. Carson, ... 38 *Ninth St.*
1856........Breed, John B., ... 404 *Broadway.*
1856........Bradish, Luther, ... 36 *East Sixteenth St.*
1856........Brodie, Alexander Oswald, . . 104 *Ninth St.*
1854....L. M. Brodhead, J. Romeyn, . . . 19*th St. and Second Avenue.*
1853........Brower, John H., ... 45 *South St.*
1853........Brown, James M., ... 20 *Waverley Place.*
1856........Brown, Stewart, ... 21 *Waverley Place.*
1856........Bull, John, ... 35 *Pine St.*
1856........Bunting, John A., . . . 11 *Livingston Place.*
1856........Burnham, Frederick G., . . . 141 *Eighth St.*
1853........Burtus, James A., . . . *Cor. Beekman and Pearl.*
1853........Butler, Charles, ... 12 *Wall St.*
1857........Butler Henry V., ... 44 *East Twenty-first St.*

1855........Camp, Hugh N., ... 101 *Wall St.*
1856........Carter, Robert, ... 5 *East Twentieth St.*
1856........Champlin, Christopher, . . 18 *Washington Place.*
1856........Champlin, Elbert H., M.D. LL.D. . 18 *Washington Place.*

RESIDENT MEMBERS.

1855........Chanler, John Winthrop, . . 6 *Wall St.*
1853........Chauncey, William, ... 10 *Old Slip.*
1856........Chilton, James R., . . . 93 *Prince St.*
1856........Churchill, Timothy G., . . . 90 *East Eighteenth St.*
1856........Clark, Albert, ... 11 *Fifth Avenue.*
1856........Clark, Frederick H., . . . 28 *University Place.*
1857........Clarke, George W., ... 218 *Fourth St.*
1856........Clay, George, ... 56 *Clinton Place.*
1856........Cobb, George T., ... 28*th St. and Fifth Avenue.*
1852........Coddington, Thomas B., . . 76 *Broad St.*
1856........Coffin, Edmund, ... 30 *Broad St.*
1856........Coit, Henry A., ... 55 *Clinton Place.*
1856........Colgate, Chas. C., ... 6 *Dutch St.*
1856........Colgate, Robert, ... 5 *West Fourteenth St.*
1856........Codwise, David, ... 27 *St. Mark's Place.*
1856........Coleman, William T., . . . 88 *Wall St.*
1855........Coles, Jeremiah.
1853........Collins, Edward K., . . . 56 *Wall St.*
1854........Colton, George W., ... 172 *William St.*
1852....L. M. Colton, Joseph H., . . . 172 *William St.*
1852........Congdon, Charles, ... 28 *Cliff St.*
1855........Conkling, Frederick A., . . 103 *Tenth* (31 *Warren*) *St.*
1856........Conner, James, ... 29 *Beekman St.*
1856........Cook, John, Jr., ... 28 *Exchange Place.*
1856........Cooley, James E., ... 78 *Fifth Avenue.*
1856........Cooper, Edward, ... 17 *Burling Slip.*
1855........Cooper, Peter, ... 17 *Burling Slip.*
1856........Cornell, Wm. W., ... 139 *Centre St.*
1856........Corning, Jasper.
1852........Cotheal, Alexander I., . . . 49 *Water St.*
1856........Cousland, William, ... 154 *West Twenty-Fifth St.*
1856........Crerar, John, ... 45 *Cliff St.*
1852........Crooks, Ramsay, ... 58 *Broad St.*
1856........Crooks, Ramsay, Jr., . . . 14 *St. Marks Place.*
1856........Crowen, Thomas J., ... 699 *Broadway.*
1854........Culbert, John W., . . . 24 *Old Slip.*
1857........Cunard, Edward, ... 4 *Bowling Green.*
1856........Currie, Chas. P., ... 33 *East Twelfth St.*
1856........Curtis, George William, . . . 27 *Washington Place.*
1857........Curtis, James L., ... 13 *East Twenty-eighth St.*
1856........Curtis, Lewis, ... 11 *East Fourteenth St.*
1856........Curtis, William E., . . . 106 *Broadway.*
1854........Curtiss, Cyrus, ... 80 *Ninth St.*
1852........Cutting, Fulton, ... 55 *Merchants' Exchange.*

RESIDENT MEMBERS.

1856........Dabney, Charles H., ... 281 *Fifth Avenue.*
1855........Daly, Chas. P., ... 84 *Eighth St.*
1855........Dambmann, Chas. F., ... 14 *Park Place.*
1852........Davis, Charles, ... 78 *Merchants' Exchange.*
1857........Davis, Samuel, ... 10 *East Twenty-fifth St.*
1856........Day, Horace H., ... 120 *Tenth St.*
1856........De Angelis, Gideon, ... 42 *West Twenty-Seventh St.*
1856........Degen, Chas. R., ... 7 *Gramercy Park.*
1856........Delano, Joseph C., ... *New Bedford, Mass.*
1856........Delano, Warren, ... 39 *La Fayette Place.*
1856........Deming, Richard T., ... 36 *Wall St.*
1856........Denison, Rev. Samuel D., . . . *Bible House.*
1856........Derby, James C., ... 119 *Nassau St.*
1856........Detmold, William, M.D., ... 103 *Ninth St.*
1856........Dillon, Robert, ... 28 *and* 30 *Burling Slip.*
1853........Dinsmore, Curran, ... 9 *Spruce St.*
1856........Dinsmore, Samuel P., ... *S. E. cor.* 14*th St. &* 7*th Av.*
1852........Disturnell, John, ... 16 *Beekman St.*
1856........Dodge, William E., ... 43 *West Thirty-First St.*
1856........Dodge, William E., Jr., ... 43 *West Thirty-First St.*
1856........Doremus, R. Ogden, M.D., . . . 70 *Union Place.*
1856........Douglass, Andrew E., ... 89 *Wall St.*
1856........Douglas, J. Hancock, ... 12 *Clinton Place.*
1856........Duncan, Alexander W., ... 2 *Washington Square.*
1856........Duncan, William Butler, . . . 233 *Fifth Avenue.*
1856........Dunnel, Henry G., M. D., ... 51 *St. Mark's Place.*
1855........Dunshee, Henry W., ... 36 *Amos St.*
1857........Dustan, John F., ... 38 *Tenth St.*

1857........Edey, Charles C., ... 105 *Wall St.*
1856........Edgar, James A., ... 95 *Front St.*
1852........Elliott, Samuel M., M. D., . . . 371 *Fourth Avenue.*
1856........Ellis, George, ... 15 *Nassau St.*
1856........Ellis, Samuel C., M.D., ... 179 *Second Avenue.*
1856........Elphinstone, William H., ... 112 *East Twenty-First St.*
1856........Ely, Abner L., ... 31 *Pine St.*
1854........Ely, Charles, ... 15 *Nassau St.*
1856........Elsworth, Henry, ... 143 *West Fourteenth St.*
1856........Embury, Daniel, ... 74 *Clark St., Brooklyn.*
1856........Emmet, Robert, ... 122 *Madison Avenue.*
1856........Emmet, Thomas Addis, ... 24 *University Place.*
1853........Eyre, Henry S. P., ... 74 *Broad St.*

RESIDENT MEMBERS.

Year	Name	Address
1856........	Faile, Edward G.,	192 *Chambers St.*
1856........	Fairchild, Ezra R., D.D.,	156 *Chambers St.*
1853........	Fearing, Daniel B.,	*5th Av. cor. of Fifteenth St.*
1856........	Fellows, Chas. H.,	74 *Beekman St.*
1856........	Fellows, Richard C.,	103 *Second Avenue.*
1856........	Felt, Willard,	14 *Maiden Lane.*
1856........	Felt, William N.,	21 *Bond St.*
1856........	Ferguson, John,	35 *Pine St.*
1856........	Fernbach, Henry,	101 *Third Avenue.*
1856........	Fetridge, William P.,	*Everett House.*
1856....L. M.	Field, Benjamin H.,	21 *East Twenty-Sixth St.*
1852....L. M.	Field, Hixon W.,	*St. Germain Hotel.*
1854........	Field, Cyrus W.,	11 *Cliff St.*
1856........	Field, David Dudley,	86 *East Twenty-First St.*
1856........	Field, Dudley,	86 *East Twenty-First St.*
1852........	Field, Maunsel B.	
1856........	Fiedler, Ernst,	32 *Broadway.*
1854........	Fish, Hamilton,	*Cor. of 17th St. and 2d Ave*
1857........	Fischel, H., D. D.,	5 *Carroll Place.*
1855........	Fisher, Richard S., M.D.,	133 *Eighth St.*
1856........	Fitch, S. S., M.D.,	714 *Broadway.*
1852........	Fleming, Augustus,	10 *Bond St.*
1856....L. M.	Folsom, George,	212 *Second Avenue.*
1856........	Ford, James K.,	115 *Wall St.*
1857........	Forrester James C.,	202 *Bleecker St.*
1856........	Forrester, Hiram M.,	134 *West Twenty-First St.*
1855........	Fosdick, William W.,	347 *Broadway.*
1856........	Foster, Frederick G.,	65 *South St.*
1856........	Fowler, Isaac V.,	*New York Hotel.*
1856........	Fowler, Frederick R.,	11 *East Twenty-Sixth St.*
1853........	Francis, John W., M.D.,	1 *Bond St.*
1856........	Fraser, Thomas,	20 *West Twenty-Second St.*
1854........	Frazer, Maj. W. D., U. S. A.,	70 *Broadway.*
1853........	French, Benjamin F.,	94 *Clinton Place.*
1856........	French, James,	153 *West Twenty-Second St.*
1855........	Frost, Joseph B.,	40 *Water St.*
1856........	Frye, Jed,	72 *West Thirty-First St.*
1856........	Fuller, Dudley B.,	42 *West Fourteenth St.*
1856........	Furniss, William,	35 *Wall St.*

Year	Name	Address
1853........	Gaillard, Joseph,	34 *South St.*
1856........	Gale, William,	28 *West Twenty-First St.*
1856........	Gans, Meyer,	52 *Exchange Place.*
1856........	Gescheidt, Anthony, Jr., M. D.	44 *Clinton Place.*

1856........Gibson, Ebenezer, ... 143 *Front St.*
1856........Gilbert, Albert, ... 146 *Grand St.*
1857........Gillette, Abraham D., D. D.. . . 29 *West Twenty-third St.*
1852........Goodhue, Robert C., . . . 64 *South St.*
1856........Gourlie, John H., ... 1 *Hanover St.*
1856........Grant, Oliver D. F., ... 11 *Gramercy Park.*
1854........Grant, S. Hastings, . . . *Mercantile Library.*
1852........Gregory, Lewis, ... 33 *Pine St.*
1857........Green, Horace, M. D., . . . *New York Hotel.*
1855........Green, James, ... 14 *Wall St.*
1857........Green, John W., M. D., . . . *Thirteenth St.*
1856........Greenwood, Isaac J., . . . 142 *West Fourteenth St.*
1853........Grinnell, Cornelius, . . . 87 *South St.*
1852........Grinnell, Henry, ... 87 *South St.*
1853........Grinnell, Moses H., . . . 78 *South St.*
1857........Grinnell, Wm. M., ... 17 *Bond St.*
1856........Griswold, Almon W., . . . 13 *East Thirtieth St.*
1856........Groshon, John, ... 73 *Sixth Avenue.*
1856........Guernsey, Egbert,. . . . 19 *West Twenty-Second St.*
1857........Gunn, Alexander, M. D., . . . 132 *Fourth St.*

1857........Haight, Rev. Benj. I., . . . 56 *West Twenty-sixth St.*
1856........Haight, Richard K., . . . 170 *Water St.*
1856........Hall, Valentine G., ... 16 *Gramercy Park.*
1856........Halstead, Caleb O., . . . 155 *Second Avenue.*
1856........Halstead, Robert, ... 1 *West Fourteenth St.*
........Hamilton, Edmund, . . . 63 *Wall St.*
1857........Hamilton, George F., . . . 40 *Broadway.*
1856........Hand, John H., ... 49 *Wall St.*
1854........Hart, M. G., ... *Washington, D. C.*
1852........Hargous, Peter A., . . . 76 *Fifth Avenue.*
1856........Harrington, Luke, ... 122 *Water St.*
1857........Harris, Elisha, M. D., . . . *Quarantine, Staten Island.*
1855........Harrison, Thomas F., . . . 41 *Greenwich Avenue.*
1854........Hasbrook, A. B., ... 193 *Twelfth St.*
1856........Hatfield, Amos F., ... 78 *East Nineteenth St.*
1856........Hawes, William E., . . . *People's Bank, Canal St.*
1852........Hawks, Rev. Francis L., D.D., LL.D., 64 *East Twenty-First St.*
1856........Haydock, George G., . . . 30 *Broad St.*
1855........Hearn, George A., . . . 425 *Broadway.*
1856........Hedden, Josiah, ... 59 *Broadway.*
1857........Hedges, Timothy, ... 19 *West Twentieth St.*
1856........Heiser, Henry A., ... *Fifth Avenue and* 36*th St.*
1856........Herrick, John J., ... 20 *West Twentieth St.*

RESIDENT MEMBERS.

1856........Herring, Silas C., ... 102 *Fifth Avenue.*
1855........Hewitt, Abram S., ... 17 *Burling Slip.*
1857........Higbee, Rev. Edward Y., ... 19 *West Twenty-sixth St.*
1856........Hodges, John T., ... 271 *Broadway.*
1856........Hoge, William, ... 40 *Wall St.*
1856........Holmes, Silas, ... 48 *Bond St.*
1856........Homans, Sheppard, ... 111 *Broadway.*
1856........Howard, Joseph, Jr., ... 34 *Broadway.*
1856........Howland, Williams, ... 106 *Wall St.*
1853........Hunt, Thomas, ... 92 *William St.*
1856........Hunt, Wilson G., ... 36 *Park Place.*
1856........Hunter, James, ... 174 *Front St.*

1852....L.M. Jay, John, ... 20 *Nassau St.*
1856........James, Albert S., ... 10 *Wall St.*
1856........Jarvis, Jay, ... 7 *Forsyth St.*
1856........Jaudon, Samuel, ... 54 *Wall St.*
1856........Jessup, Morris K., ... 44 *Exchange Place.*
1856........Johnson, Amos, ... 73 *East Twelfth St.*
1855........Johnson, Bradish, ... 64 *Fifth Avenue.*
1856........Johnson, Hezron A., ... 105 *Fifth Avenue.*
1855........Johnston, John T., ... 69 *Wall St.*
1856........Johnston, James B., ... 56 *Tenth St.*
1852........Jones, John D., ... 51 *Wall St.*
1855........Jones, Rev. Lot, ... 77 *Second Avenue.*
1857........Jones, Lewis C., ... 24 *West Sixteenth St.*
1856........Joslin, Benjamin F., M. D., . . . 72 *University Place.*
1857........Judson, Stiles W., ... 113 *Ninth St.*

1855........Kearny, Edward, ... 139 *Front St.*
1856........Kelly, John J., ... 137 *Front St.*
1856........Kelly, Robert E., ... 45 *Beaver St.*
1854........Kennedy, R. Lenox, ... 63 *Beaver St.*
1856........Ketchum, Hiram, Jr., ... 29 *William St.*
1853........Kimball, Richard B., ... 49 *Wall St.*
1854........King, Charles, LL. D., ... *Columbia College.*
1852........Kingsland, Ambrose C., . . . 55 *Broad St.*
1856........Kirkwood, James P., ... 353 *Fulton St., Brooklyn.*
1856........Kissam, Richard S., M. D., . . . 9 *Great Jones St.*
1853....L. M. Knapp, Shepherd, ... 33 *Wall St.*
1856........Kneeland, Charles, ... 49 *William St.*
1856........Knevitt, George M., ... 65 *Wall St.*
1856........Knoedler, M., ... 366 *Broadway.*

RESIDENT MEMBERS.

1856........Lamson, Charles, ... 114 *Tenth St.*
1856........Lane, Robert, ... 215 *Tenth St.*
1856........Lanier, James F. D., . . . 54 *Tenth St.*
1852........Latson, John W., ... 79 *Nassau St.*
1854........Lawrence, Alex. M., . . . 82 *East Twenty-First St.*
1856........Lawrence, William S., . . . 94 *Broad St.*
1852....L.M. Leavitt, Joshua, ... 22 *Beekman St.*
1856........Leake, Godfrey W., . . . 142 *Canal St.*
1856........Leet, Allen N., ... 135 *Water St.*
1854........Lefferts, Marshall, . . . 70 *Broad St.*
1856........Leroy, Jacob R., ... 26 *La Fayette Place.*
1856........Lindsley, Caleb F., . . . 11 *Wall St.*
1856........Little, Edward B., ... 20 *Washington Place.*
1852....L.M. Livingston, Cambridge, . . 17 *Wall St.*
1857........Livingston, Mortimer, . . . 53 *Broadway.*
1857........Locke, Richd. A., ... *Staten Island.*
1856........Lockwood, Le Grand, . . . 127 *West Fourteenth St.*
1857........Loomis, Hezekiah B., . . . 30 *University Place.*
1857........Lord, James Cooper, . . . 24 *Gramercy Park.*
1857........Low, Abiel A., ... 31 *Burling Slip.*
1852........Lowry, Francis B., ... 156 *Broadway.*
1856........Ludlow, Ezra, ... 43 *Water St.*
1857........Lyman, Samuel P., ... *Astor House.*

1856........Manners, David S., . . . 7 *Wall St.*
1856........Marsh, Samuel, ... 45 *John St.*
1856........Marshall, Charles H., . . . 38 *East Fourteenth St.*
1856........Mason, Daniel G., ... 110 *Duane St.*
1857........Mason, John, ... 57 *Henry St., Brooklyn.*
1856........Mason, John L., ... *Trinity Building.*
1856........Mason Sidney, ... 132 *Fifth Avenue.*
1856........Maxwell, John, ... 140 *West Fourteenth St.*
1857........Mayo, Wm. S., M. D. . . . 132 *Second Avenue.*
1854........McAlpin, William J., . . . *Office Erie R. R.*
1852........McKaye, James, ... 82 *Broadway.*
1856........McKillop, John, ... 5 *Beekman St.*
1856........McNamee, Theodore, . . . 112 *Broadway.*
1852........McVickar, William H., . . . 18 *Exchange Place.*
1857........Meigs, Henry D., Jr., . . . 33 *Seventh Avenue.*
1854........Merriam, E., ... *Brooklyn Heights.*
1856........Meyer, Henry, ... 30 *Cliff St.*
1852........Myers, T. Bailey, ... 229 *Broadway.*
1855........Miles, Pliny.
1856........Miller, Edmund H., . . . 63 *Wall St.*

RESIDENT MEMBERS.

1856........Miner, William, M. D., ... 65 *Irving Place.*
1857........Minturn, James B., ... 29 *Burling Slip.*
1856........Minturn, Robert B., ... 60 *Fifth Avenue.*
1856........Minturn, Thomas R., ... 115 *Pearl St.*
1854........Mitchell, Matthew, ... 111 *Water St.*
1856........Mitchell, Roland G., ... 8 *West Nineteenth St.*
1856........Mitchill, Samuel L., ... 25 *Union Place.*
1856........Monroe, Ebenezer, ... 247 *Broadway.*
1854........Moran, Charles, ... 1 *Hanover Square.*
1854........Morgan, Capt. E. E., ... 70 *South St.*
1856........Morgan, Edwin D., ... 68 *Front St.*
1853........Moore, George H., ... *N. Y. Historical Society.*
1856........Morse, A. Wellman, ... 63 *Wall St.*
1854........Morse, Sidney E., ... 138 *Nassau St.*
1855........Morton, Alvin C., ... 110 *Broadway.*
1856........Mount, Alfred R., ... 64 *Wall St.*
1857........Muhlenberg, Rev. W., ... *Cor. Twentieth St. & Sixth Av.*
1856........Murdock, Uriel A., ... 291 *Fifth Avenue.*

1857........Nash, Alanson, ... 36 *Beekman St.*
1856........Negus, Thomas S., ... 100 *Wall St.*
1856........Neilson, Anthony B., ... 31 *East Twenty-second St.*
1856........Nestell, Daniel D. T., M. D., . . . 68 *West Thirty-second St.*
1856........Niblo, William, ... *Niblo's Garden.*
1856........Nichols, Effingham H., ... 377 *Fourth St.*
1856........Nichols, William S., ... *Cor. Ex. Place & William St.*
1856........Nicolay, Albert H., ... 4 *Broad St.*
1856........Norton, Robert, ... 9 *South St.*
1856........Noyes, William Curtis, ... 50 *Wall St.*

1856........Oakley, Edward J., ... 133 *West Twenty-third St.*
1855........Oakley, Henry A., ... 66 *Wall st.*
1856........O'Brien, James, ... 150 *Broadway.*
1856........Odell, George F., ...
1856........Oelrichs, Edwin A., ... 68 *Broad St.*
1857........Opdyke, George, ... 79 *Fifth Avenue.*
1852........Osgood, A. A.
1855........Osgood, Rev. Samuel, ... 86 *West Eleventh St.*

1856........Palmer, John J., ... 230 *Fourth St.*
1856........Parker, Edward H., M. D., . . . 279 *Fourth Avenue.*
1856........Parmly Ehrick, ... 1 *Bond St.*
1856........Pavenstedt, Edmund, ... 38 *New St.*

1855........Pell, John Augustus.
1856........Penfold, Edmund, ... 178 *Front St.*
1856........Pentz, Benj. J., ... 153 *Bowery.*
1856........Perit, Pelatiah, ... 64 *South St.*
1856........Phillips, James W., ... 52 *South St.*
1855........Pierrepont, Edwards, ... 27 *Wall St.*
1852....L. M. Pierrepont, Henry E., ... 53 *Broadway.*
1856........Pierson, Henry L., ... 24 *Broadway.*
1852....L. M. Poor, Henry V., ... 9 *Spruce St.*
1856........Potter, Gilbert, Jr., ... 181 *Front St.*
1856........Porter, Mortimer, ... 110 *Ninth St.*
1856........Powers, Thomas J., ... 50 *Wall St.*
1853........Pratt, George W., ... 92 *Fifth Avenue.*
1853........Pratt, Zadoc, ... 92 *Fifth Avenue.*
1852........Prime, Frederick, ... 54 *Wall St.*
1855........Pullen, John A., ... 221 *West Fourteenth St.*
1856........Putnam, C. S.. . . . . . . 35 *Bond St.*
1857........Pine, Percy R., ... 25 *East Twenty-second St.*

1856........Rait, Robert, ... 154 *East Nineteenth St.*
1856........Randolph, Anson D. F., . . . 64 *Bedford St.*
1856........Rankin, Robert G., ... *Trinity Building.*
1856....L. M. Raphall, Rev. Morris J., DR. P., . 94 *Macdougal St.*
1857........Reid, John W., ... 11 *Old Slip.*
1856........Remsen, William, ... 22 *William St.*
1856........Richards, Joseph W., M. D., . . 12 *Clinton Place.*
1856........Riker, John H., ... 150 *Nassau St.*
1856........Robbins, Henry A., ... 15 *Maiden Lane.*
1855........Robertson, Daniel A., ... 34 *Barclay St.*
1853....L. M. Robinson, Rev. Edward, D. D., . . 257 *Greene St.*
1853........Robinson, Henry B., ... 111 *Broadway.*
1856........Robbins, George S., ... 54 *William St.*
1856........Rockwell, Charles W., . . . 8 *Wall St.*
1856........Root, Russell C., ... 16 *Nassau St.*
1856........Rowe, Jacob, ... 8 *Broadway.*
1854........Ruggles, Samuel B., ... 110 *Broadway.*
1852....L. M. Russell, Archibald, ... 45 *Tenth St.*
1856........Russell, Charles H., ... 2 *Great Jones St.*
1856........Russell, William H., ... 686 *Broadway.*
1854........Rutherfurd, Lewis M., ... 175 *Second Avenue.*
1855........Rutherfurd, Robert W., . . . *Belleville, N. J.*

1856........Salmon, James, ... 10 *Livingston Place.*
1856........Sampson, Joseph, ... 2 *Bond St.*

RESIDENT MEMBERS.

1856........Sanger, George J.,. . . . 193 *Pearl St.*
1856........Sargent, Epes W., ... 17 *Broadway.*
1856........Schieffelin, Sidney A., . . . 13 *East Twenty-sixth St.*
1856........Schmidt, John W., M. D., . . . 5 *East Twelfth St.*
1856........Schermerhorn, William C., . . 68 *Wall St.*
1856........Schreiner, Osmond H., . . . 71 *West Nineteenth St.*
1856........Schroeter, George, . . . 111 *Broadway.*
1856........Schuchardt, Frederick, . . . 29 *Washington Place.*
1857........Seabury, Rev. Samuel, . . . 125 *West Thirteenth St.*
1856........Sears, Alfred.
1856........Sears, Herman B., ... 71 *West Twenty-second St.*
1856........Sedgwick, Theodore, . . . 44 *Wall St.*
1856........Seton, Alfred, ... 49 *Wall St.*
1854........Sewall, Henry F., ... 78 *South.*
1856........Seymour, Isaac, ... 201 *West Fourteenth St.*
1853........Seymour, Horatio, . . . *Albany.*
1857........Seyton, Charles S., ...
1856........Shaffer, Chauncey, . . . 71 *East Twelfth St.*
1856........Shaffner, Talford P.
1856........Sheffield, Joseph B., . . . 40 *East Twenty-second St.*
1856........Sheldon, Preston, ... 92 *Market Slip.*
1856........Shepherd, Daniel,. . . . 124 *West Fourteenth St.*
1857........Sheppard, George F., . . . *St. Nicholas Hotel.*
1856........Sherman, John, ... 11 *Coenties Slip.*
1856........Sherman, Watts, ... 13 *Nassau St.*
1853........Sibell, William E., . . . 20 *Wall St.*
1857........Simes, John D., ... 10 *Lexington Avenue.*
1857........Smith, Edwin, ... 56 *East Nineteenth St.*
1852........Smith, J. Calvin, ... 71 *Nassau St.*
1853........Smith, James McC., M. D., . . 55 *West Broadway.*
1853........Smith, James O., M. D., . . . 81 *Clinton Place.*
1856........Smith, U. J., ... 38 *East Fourteenth St.*
1856........Smith, Capt. Melancthon, U. S. N., . 2 *West Twenty-fifth St.*
1856........Spencer, Lorillard, . . . *Cor. Fifth Av. & Sixteenth St.*
1856........Spofford, Paul N., ... 4 *East Fourteenth St.*
1855........Squier, E. George, . . . 16 *St. Mark's Place.*
1853........Stevens, Rev. Abel, ... 200 *Mulberry St.*
1854........Stevens, Alexander H., M. D.,. . 6 *Lafayette Place.*
1853........Satterthwaite, T. B., . . . 50 *Wall St.*
1856........Stebbins, Henry G., . . . '2 *West Sixteenth St.*
1856........Stokes, Benaiah G., ... 46 *West Twenty-fifth St.*
1856........Storm, Walter, ... 104 *Broad St.*
1853........Stout, Aquila G., ... 71 *Wall St.*
1856........Stout, Theodore, ... *St. Denis Hotel.*

RESIDENT MEMBERS.

1856........Stratton, James T., ... 111 *Fourth Avenue.*
1856........Stringer, James, ... 429 *Broome St.*
1856........Strong, William K., ... 40 *Union Place.*
1856........Strube, Frederick, M. D., . . . 4 *Leroy Place.*
1857........Stucken, Charles, ... 76 *Beaver St.*
1856........Sturgis, Russell, ... 164 *Tenth St.*
1856........Stuyvesant, Gerard, ... 124 *Second Avenue.*
1855........Stuart, Robert L., ... 171 *Chambers St.*
1856........Sutphen, Ten Eyck, ... 404 *Broadway.*
1856........Suydam, Peter M., ... 25 *Waverley Place.*

1854........Taber, Charles C.,. . . . . 76 *Wall St.*
1856........Talmage, Daniel, ... 60 *Water St.*
1855........Taylor, George.
1857........Taylor, Isaac E., M. D., ... 33 *East Sixteenth St.*
1856........Taylor, Rev. Thomas House, D. D., *Grace Church, Broadway.*
1855........Tellkampf, Theodore A., M. D., . . 198 *Fourth St.*
1856........Thomas, Griffith, ... 101 *Fifth Avenue.*
1857........Thompson, David, ... 25 *Lafayette Place.*
1854........Thompson, Rev. Joseph P., . . . 63 *Amity St.*
1856........Thomson, John, ... 232 *West Twenty-second St.*
1856........Thurston, Frederick G., . . . 94 *South St.*
1856........Tibbetts, Henry B., ... 111 *Broadway.*
1856........Tiffany, Charles F., ... 195 *Tenth St.*
1856........Tillou, Francis R., ... 42 *Laight St.*
1856........Timpson, R. H., ... 161 *Front St.*
1854........Tileston, Thomas, ... 29 *Broadway.*
1852........Tomes, Benjamin, ... 6 *Maiden Lane.*
1855........Tomes, Francis, Jr., ... 12 *West Twenty-first St.*
1856........Tooker, Richard A., ... *Nassau Bank.*
1857........Townsend, Dwight, ... 102 *Wall St.*
1857........Townsend, Effingham, ... 113 *Broadway.*
1856........Townsend, Randolph W., ... 140 *Broadway.*
1856........Townsend, William A., . . . *Cor. Broadway & Ann St.*
1856........Travers, William R., ... 20 *William St.*
1856........Treadwell, Henry R., ... 550 *Broadway.*
1856........Tremain, Edwin R., ... 130 *East Thirteenth St.*
1856........Trenor, John, ... 1 *University Place.*

1856........Vandervoort, Peter H., ... 243 *Water St.*
1857........Van Blarcom, ... 57 *Pine St.*
1856........Van Buren, Wm. H., ... 24 *East Twenty-second St.*
1856........Vermilye, Wm. E., ... 44 *Wall St.*

RESIDENT MEMBERS.

1854........Viele, Egbert L., ... 27 *Amity St.*
1856........Vincent, Edward, ... 106 *Bleecker St.*
1854........Vinton, Rev. Francis D., D. D., . . 31 *Vesey St.*

1856........Wade, John, ... 110 *Broad St.*
1854........Waddell, Wm. C. H., ... 24 *William St.*
1856........Walke, Cornelius, M. D., ... 213 *Second Avenue.*
1856........Walker, Francis C., ... 43 *West Nineteenth St.*
1856........Waterbury, Lawrence, ... 104 *East Fifteenth St.*
1856........Warren, Richard, ... 9 *Clinton Place.*
1856........Warren, James K., ... 65 *Wall St.*
1854........Ward, Augustus H. . . . . 14 *Washington Place.*
1857........Ward, Rev. Henry Dana, ... 38 *West Twenty-third St.*
1853....L. M. Watts, Ridley, ... 27 *East Twentieth St.*
1854........Watts, Robert, M. D., ... 200 *Twelfth St.*
1854........Webb, William H., ... 72 *St. Mark's Place.*
1856........Weed, Ebenezer S., ... 29 *Ferry St.*
1856........Wells, Jacob, ... 52 *John St.*
1856........Westermann, Bernard, ... 290 *Broadway.*
1856........Wetmore, David, ... 105 *Ninth St.*
1854........Wetmore, Samuel, ... 73 *South St.*
1855........Wheeler, David E., ... 237 *Broadway.*
1856........Whipple, Augustus W., ... 24 *Merchants' Exchange.*
1856........Wheatley, Charles M., ... 9 *Bleecker St.*
1856........White, John H., ... 59 *Nassau St.*
1856........White, Ezra, ... 18 *West Sixteenth St.*
1856........Whitlock, Benjamin M., ... 13 *Beekman St.*
1856........Williams, John E., ... *Metropolitan Bank.*
1856........Wilder, Benjamin G., ... 122 *Water St.*
1857........Wilson, Harris, ... 8 *Wall St.*
1855........Winthrop, Benj. R., ... 134 *Second Avenue.*
1854....L. M. Witthaus, Gustavus H., . . . 34 *Barclay St.*
1854....L. M. Witthaus, Rudolph A., ... 34 *Barclay St.*
1856........Wood, Fernando, ... *City Hall.*
1856........Wood, Rev. A. Augustus, ... 21 *Amity Place.*
1855........Wright, William W., ... 93 *Liberty St.*
1852........Wyman, Luther B., ... 38 *Burling Slip.*

1856........Young, Francis G., ... 59 *Nassau St.*

1852....L. M. Zimmerman,* John C., ... 74 *South St.*

---

* Deceased.

www.ingramcontent.com/pod-product-compliance
Lightning Source LLC
LaVergne TN
LVHW011126110826
845150LV00008B/2263
* 9 7 8 1 4 1 8 1 9 5 5 0 2 *